contolled operations to seperate noble metals (gold, silver and platinum) from base metals (lead, copper, zinc, aresenic, antimony and bismuth) present in the ore. The process is based on the principle that precious metals do not oxidize or react chemically, unlike base metals; So when they are heated at high temperatures, the precious metals remain apart and the others react forming slag and other compounds.

ABBREVIATIONS COMMON TO THESE PROCEDURES--

AC- ammonium chloride

AR- aqua-regia

SN- sodium nitrate

SC- stannous cloride

NA- nitric acid

CLS- cyanide leach substitute

SMB- sodium meta-bisulfite

ZP- zinc powder

HA- hydrochloric acid

PDS- potassium dichromate salts

SECTION 1

SAFETY FIRST,always

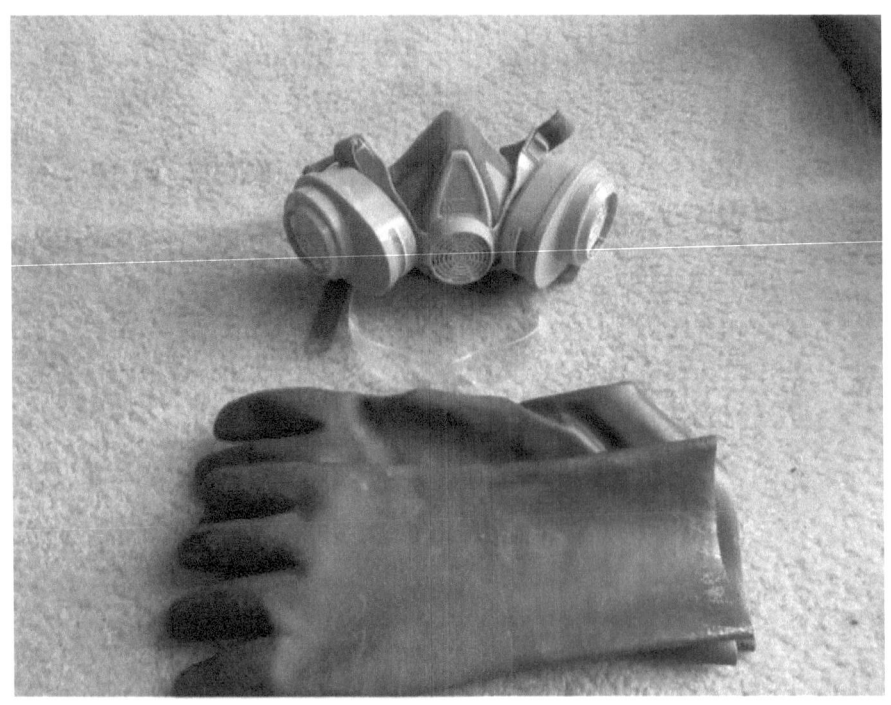

In this process you will be working with basic household chemicals. By themselves theses chemicals are fairly harmless. The combination of such chemicals which will be presented in this process can be EXTREMELY HAZARDOUS; By such there

PREFACE

Hello, my name is Rev. Nicholas w. Nickerson, and for years 1 watched myself and other people 1 know throw away broken cell phones, VCR, DVD players and all sorts of electronic devices. Do you realize the amount of gold, silver and platinum thats thrown away each year? Well folks 1 do, So 1 started researching ways to salvage those precious metals.

Folks, outlined in this book will be all the research my brother and 1 collected over the course of 3 years. The methods in this book are for the common person to follow. Althought you wont get filthy rich doing this, you will be able to rnd up with a healthy nest egg.

Remember that playing with and mixing chemicals is very DANGEROUS and can kill you. So please, follow all safety guide lines 1 state. 1 hope you find the informaion in this book helpful, be safe and happy refining in the modern world.

I will detail a SAFE process of turning old electronic parts into little round lumps of gold. There are many tutorials out there that claim to offer methods of gold recovery such as "cupellation"*. These processes are DANGEROUS andput the user at risk of inhaling vaporized metals.

No over-the-counter breather will protect you from these dangers; Only lab grade isolated breathing systems will protect against metal vapors.

SEE HEALTH HAZARDS ON STEP 6

a side cautionary note, this process is not suitable to refine panned or karat (gold alloyed with silver) gold. This is best suiuted for gold-plated and pure gold objects.

THIS SHOULD BE NEVER ATTEMPTED INDOORS

What is cupellation?

Cupellation is a refining process in metallurgy, wheres ores or alloyed metals are treated under high temperatures and are a couple of safety items which are not optional. The following items are required, failure to obtain these items beforehand can possibly compromise your health

Item #1. Breathing Equipment

You will need a chemical mask rated at least a p100. If you do not know what a p100 rating is, it is your responsibility to find out.

Item #2. Heavy Duty Gloves

Pvc or rubber will work just fine, do not use latex or nitrile.

Item #3. Safety Glasses

Any pair will work as long as there safety rated not solvent rated.

EQUIPMENT NEEDED

Helpful items which will make doing these experiments easier.

1. Borosil glass labware

2. 1-2.5 gallon bucket

3. 1-1 gallon bucket, with holes drilled in the bottom, to be used as a strainer

4. 60 ounce pickle jar

5. coffee filters

6. filter holder from a coffee machine

7. spray bottle

8. digital scale, able to measure in grams and ounces

9. pipettes and test tubes

CHEMICALS NEEDED

These are the most widely available---

1. clorox bleach or sodium hypochlorate at 6%

2. muriatic acid 34% HCL

3. hydrogen peroxide 3% but no greater than 5%

4. stumpout by BONIDE- sodium bisulfite at 99.9% as specified by the MSDS or sodium meta-bisulfite

You cannot substitue these chemicals with any other, doing so will not give the desired results.

SECTION 2

SELECTING MATERIALS TO BE RECOVERED

T

he number one rule in refining gold, is to concentrate the gold away from all other base metals prior to using acid or aqua-regia systems. Mechanical seperation will yeild the best results.

I was taught by the method of crap in, crap out. You want your materials to be as clean as possible with the least amount of garbage on them as possible.

Gold fingers are typically found on PCI, ISA, memory, sims and dimm cards. They create the corrosion free connection between the computer and the expansion card.

(MATERIALS- newer processors, PCB fingers and other heavy gold plated materials.

TIP- if you are using aqua-regia the spent AR can be used to pre-treat new style pentium processors with plated pins and board fingers. Be sure to neutralize the AR bath with enough urea prior to doing this.

The left-over gold hollow pins and flakes can then be refined in aqua-regia. The processor bodies can be crushed and refined in AR as a batch. There is hidden gold inside them.

MATERIALS- older processors with solid pins. Exterior pins can be removed by soaking in a nitric bath or shearing them off mechanically. The removed pins can now be refined with AR.

The heat disipation plates need to be seperated and refined in an electrolitic system or AR.

MATERIALS- fines, bench sweepings and solderd on PCB components. First treat them in a Aqueous Sodium Nitrate bath. The sodium nitrate and water will remove the solder. Seperate the chips and components.

MATERIALS- trace elements that might go into a nitric solution... If silver or palladium were present, they will desolve into the nitric acid solution. Silver and palladium can be precipitated from the waste liquid.

MATERIALS- precious metal fines, first desolve the fines in AR heating up to 120-180 degrees F will be required for palladium. Heating speeds up the overall desolving time of all metals. All fines should be desolved after a short time.

MATERIALS- lightly plated items, lightly plated items should be removed in an electrolitic system.)

This process can also be used to recover gold from CPU's such as pentium pro, cyrix or older 486 and lower generation chips.

SECTION 3

THE FIRST CHEMICAL BATH

Objective of this first chemical bath is to remove a large majority of the base metals such as nickle, zinc and copper.

After cleanly removing your plated sections of computer parts, remove any visible capacitors, transistors or other components. Be sure to remove any steel or iron parts attached, these materials can foul the reaction causing poor results.

TOOLS--

1-100 ml beaker

1-1 gallon strainer bucket

1-2.5 gallon bucket

CHEMICALS--

muriatic acid

hydrogen peroxide

materials you are looking to recover.

First, place your 1 gallon strainer bucket inside your 2.5 gallon bucket. Place your parts to be recovered inside your 1 gallon strainer bucket.

Add enough muriatic acid to cover your gold plated parts, about 1 cm above your materials. Keep track of how many ml's are added. You want to keep your waste acids to a minimum.

Second, to activate the solution, you want to add your hydrogen peroxide 3-5% solution in a 2:1 ratio. 2 parts muriatic acid 1 part hydrogen peroxide.

After the peroxide has been added to your muriatic acid the mixture should start to bubble and the base metals should begin to dissolve. At this point you should be wearing your safety

gear because a few unfriendly fumes are being released.

This process can take up to 24-48 hours, this works best at

the temperatures of 80-90 degrees F.

When this has been completed, you will see a batch of gold

foils in your now dark green acid.

SECTION 4

FILTERING THE GOLD FOILS

Items required--

coffee filters

filter holder

60 ounce pickle jar or a glass container large enough to catch

your waste acids.

Simply pour the contents of the 2.5 gallon bucket into the

pickle jar using the coffee filters to catch any gold foils floating on the surface of the acid. Once you have poured off all the acid you will be left with a pile of gold foils in your filter. Scoop these out into another container for further processing.

CONGRATULATIONS, you are about 50% finished.

HEALTH HAZARDS

Trying to melt the gold foils at this stage may lead to the following health hazards.

ZINC- inhalation of zinc oxide fumes can occur when welding or cutting on zinc-coated metals.Exposure to these fumes are known to cause metal fume fever. Symptoms of metal fume fever are very similar to those of common influenza. They include fever (rarely excceding 120 degrees F), chills, nausea, dryness of the throat, cough, fatigue and general weakness and aching of the head and body. The victim may sweat profusely for a few hours, after which the body temperature begins to return to normal. The symptoms of metal fume fever have rarely, if ever lasted beyond 24 hours. The subject can appear more susceptible to the onset of this condition on mondays or on weekdays following a hoilday than they are on other days.

BERYLLIUM- beryllium is sometimes used as a alloying element with copper and other base metals. Acute exposure to high concentrations of beryllium can result in chemical pneumonia. Long-term exposure can result in shortness of breath, chronic cough and significant weight loss accompanied by fatigue and general weakness.

MERCURY- mercury compounds are used to coat metals to prevent rust or inhibit foliage growth (marine plants). Under the intense heat of the arc or gas flame, mercury vapors will be produced. Exposure to these vapors may produce stomach pains, diarrhea, kidney damage or respiratory failure. Long-term exposure may produce tremors, emotional instability and hearing damage.

LEAD- inhalation or ingestion of lead oxide fumes and other lead compounds will cause lead poisoning. Symptoms include metallic taste in the mouth, loss of appetite, nausea, abdominal cramps and insomnia.In time anemia and general weakness (cheifly in the muscles of the wrists) develop. Lead adversely affects the brain, central nervous system, circulatory system, reproductive system, kidneys and muscles.

CADMIUM- cadmium is used frequently as a rust preventive coating on steel and as an alloying element. Acute exposure to high concentrations of cadmium fumes can produce severe lung irritation, pulmomary adema and in some cases death. Long-term exposure to low levels of cadmium in the air can result in emphysema (a disease affecting the ability of the lungs to absorb oxygen) and can damage the kidneys. Cadmium is classified by OSHA, NIOSH and the EPA as a potential human carcinogen.

DISPOSING OF YOUR WASTE ACIDS

Place your waste acids in a seperate container. I use a 2 or 3 litre bottle. Label these bottles as CLCu2 with a black sharpie, under that write may also contain trace amounts of nickle, zinc, beryllium, cadmium and lead. You can take these to a local hazard or recycle center.

Do not dump your waste acids down the drain or on the ground. It is IMPORTANT to be responsible with your wastes.

SECTION 5

CREATING STANNOUS CLORIDE

These foils are about 20% copper and 80% gold.

In order to reach a higher level of purity a second bath is required. For that we will need to create a test chemical called stannous cloride or SNCL2.

This metal is essential for identifing how much gold is left in our second bath.

For this you will need-

1 beaker

30 ml muriatic acid

1 gram of tin metal or shot

Weigh it out using your digital scale

It takes approximately 4 hours to completely dissolve the tin metal shot or instantly for the powder.

The resulting solution should be a bright yellow.

SECTION 6

THE SECOND CHEMICAL BATH- REFINING YOUR GOLD FOILS

This step should never be attempted without your chemical mask.

For this chemical bath I will be reproducing the process on a smaller scale. This is by far the most hazardous step with the most reactive wastes.

For this step I am going to use a 100 ml beaker and about 3 grams of foils.

I will be using muriatic acid and clorox bleach.

I will be adding only enough muriatic acid to cover my 3 grams of foils. Using a pipette I will slowly add clorox bleach until all the foils have desolved. You can use a glass stirring rod to help speed up the process.

After your foils have dissolved the resulting solution will be a deep golden yellow maybe even a little orange in color.

It may be desirable to filter your solution with the coffee filters one more time, depending on if little bits of green plastic from the boards found there way into this step.

SECTION 7

FILTERING PARTICLES

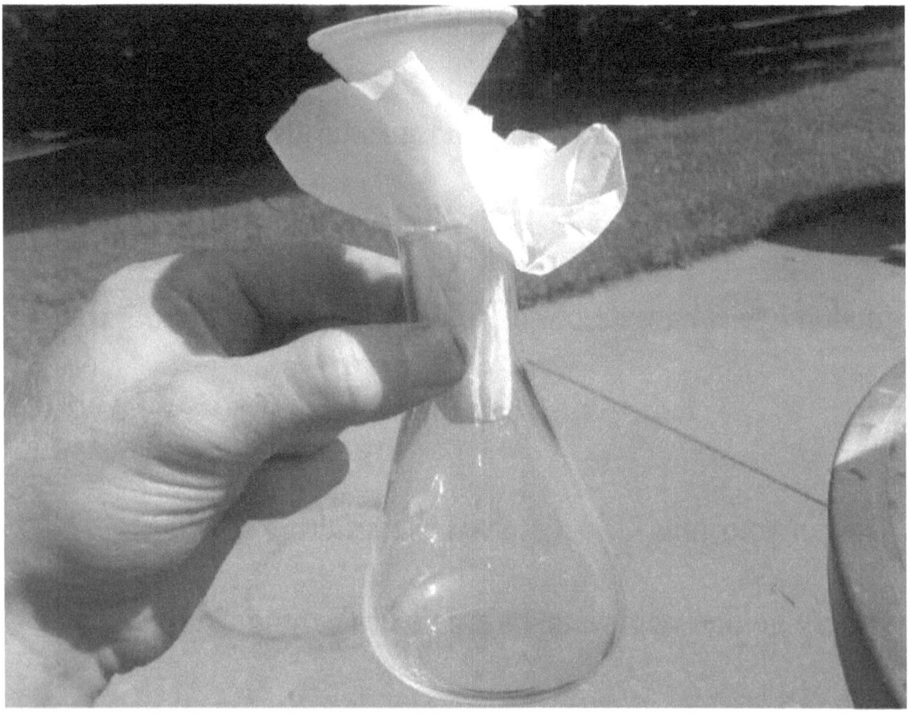

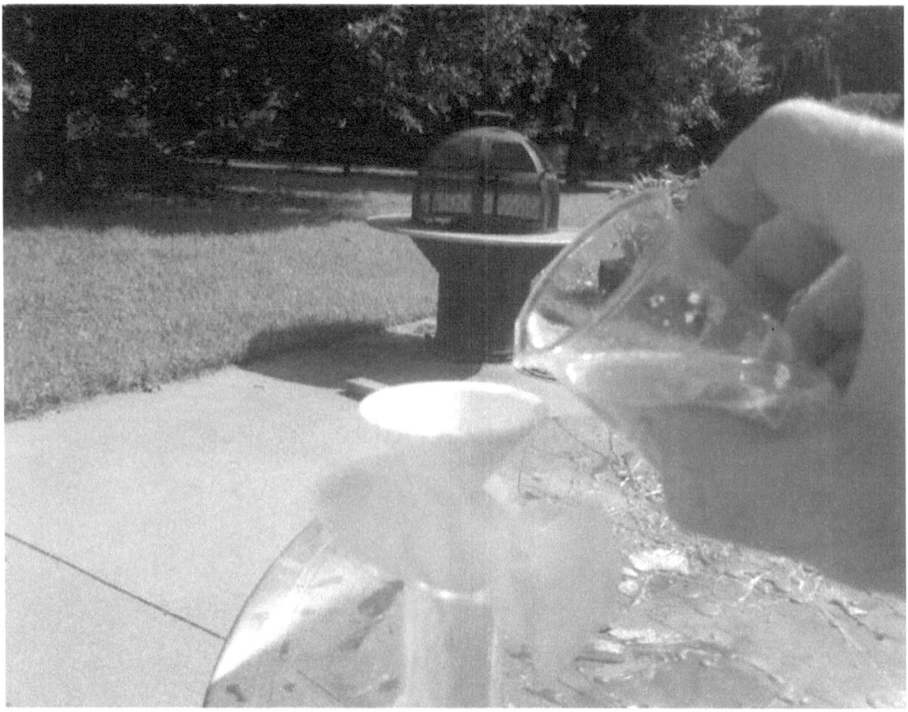

After dissolving ALL the foils you might notice a small amount of inductors, caps and transistors that wormed there way into the final process. This is normal.

Simply filter again using a long-neck funnel and a tall flask. The resulting solution should be a deep and bright golden yellow.

TESTING YOUR SOLUTION FOR GOLD

OK, this is the exciting part and one where you get to see some real chemistry at work...

you will need...

Stannous Cloride, created earlier

1 coffee filter

2 pipettes or glass eye dropper

Simply place one dot of recovered solution on a piece of filter paper. Next place one drop of your stannous cloride on the other, the resulting reaction should turn purple.

If yours did, CONGRATULATIONS you have created Auric Cloride or Aucl3, a gold-bearing solution. If your test didnt turn purple, add a little more tin to your stannous cloride and heat it gently with a alcohol burner and BOROsilicate test tube, Repeat

test.

Depending on how much gold is in your solution it may appear

black.

SECTION 8

RECOVERING YOUR GOLD

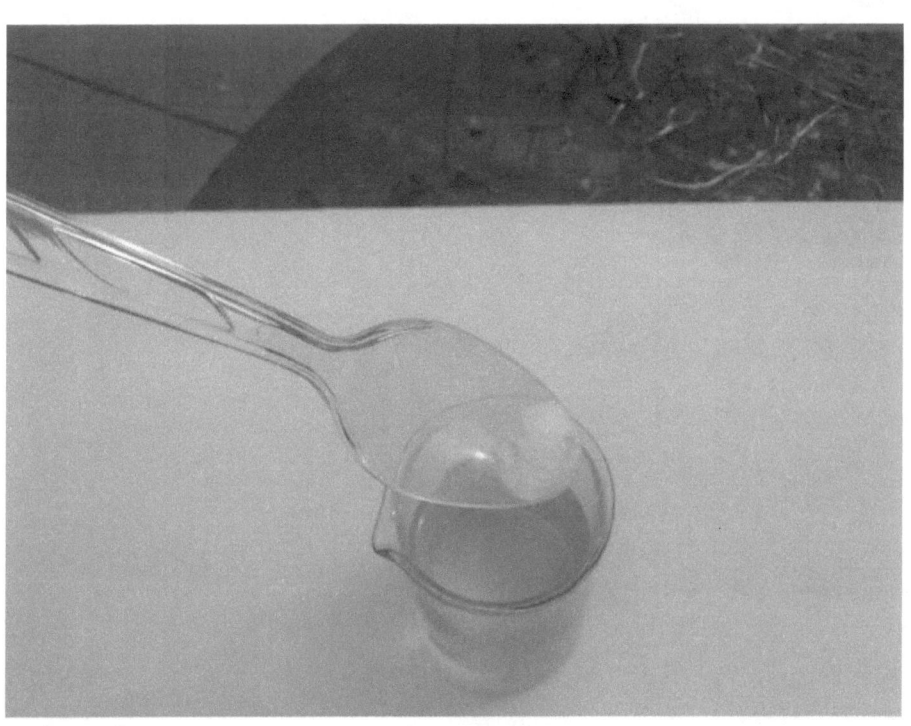

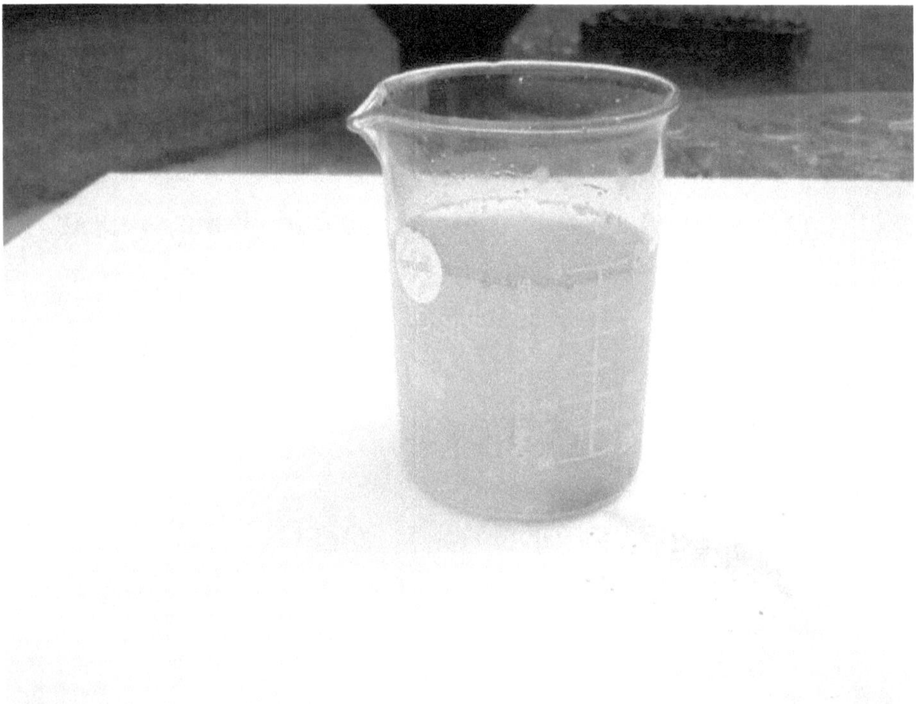

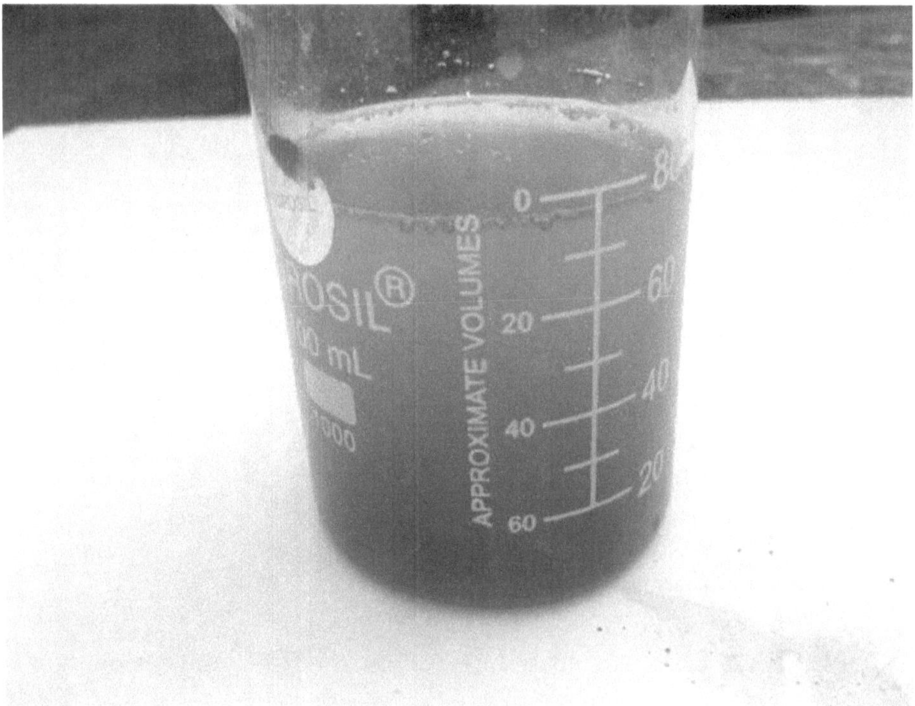
APPROXIMATE VOLUMES

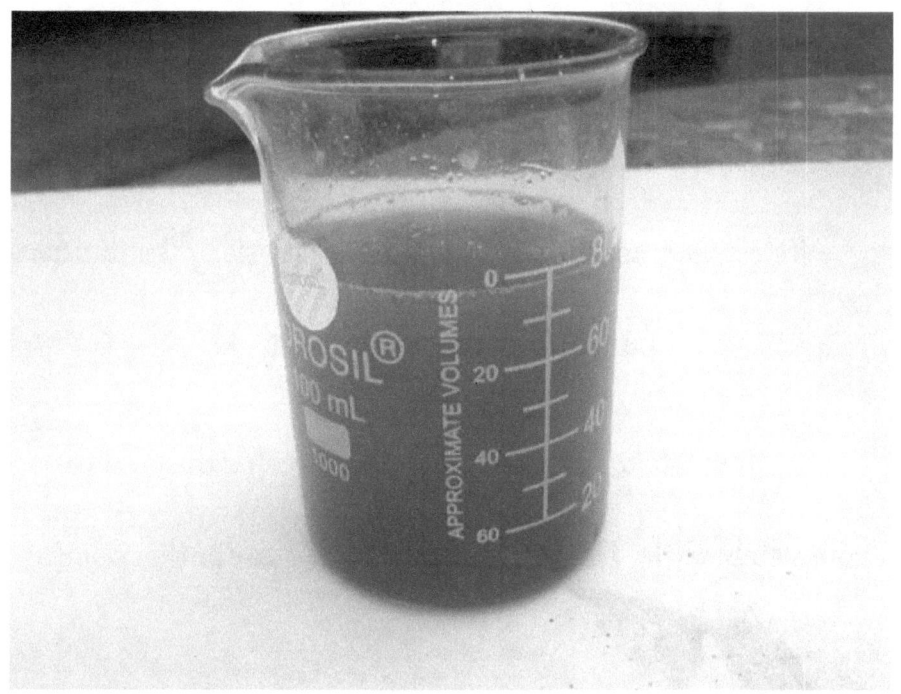

Recovering the gold in your solution is one of the coolest steps.

whats needed...

a plastic spoon

stumpout

Simply add very small amounts of stumpout,your mixture

should start to react.This is caused by the SO2 gas which helps to

chemically displace the gold in the solution.

When the reaction has finished you will have driven out

the majority of the chlorine and replaced it with sulfur. Any gold trapped in the solution has now become a brown precipitate at

the bottam of your beaker.

Retest your solution with stannous cloride, if it still turns

purple, add more stumpout.

It will take approximately 10 hours for the gold to

completely settle. The solution should be clear at this point.

THAT BROWN POWDER IS PURE GOLD!!!!

Save all your brown powder in a seperate container, I will

discuss how to melt your powder into a little gold button in a

later chapter.

CHAPTER TWO

MIXING INSTRUCTIONS AND SUPPLIES FOR OTHER

REFINING METHODS

Required supplies...

hydrochloric acid (muriatic acid)

coffee filters

paint filters

glass containers

pure distilled water (do not use treated water)

ammonia

kiln or torch (fire brick needed for torch firing)

supplies for silver...

soda ash (fluxing agent)

non-iodized salt

optional supplies...

ammonium chloride (for precipitating platinum)

zinc powder (substitute for sodium meta-bisulfite, needed for precipitating gold)

STANNOUS CLORIDE-

Although this was covered in an earlier section, heres another way to make it.

Dissolve 95% tin and 5% antimony (tin solder) into a small amount of AR.

Shelf life is 7-10 days

Used as a precious metals detection liquid.

Platinum/iridium/rhodium- deep yellow color reaction

gold- purple to black reaction color.

Auqu-regia liquid bath (for desolving gold and platinum metals).

Put one cup (approximately one pound) of sodium nitrate into a container per each quart of AR you are desiring.

Add one quart of HA for each cup of SN.

HINT- for each cup of SN, first add one cup of distlled water to quickly desolve most the SN.

I prefer to mix my AR in 4-5 gallon batches in a vented bucket. It does not matter how long it sets, it just get stronger with time. After 48 hours its strong enough to use.

Sodium Nitrate (nitrogen) bath (for desolving iron, copper and other unwanted metals)

One cup SN, one cup distilled water, for soaking processors, gold fingers, pins and other plated materials (this liquid will dissolve solder great other metals not so much).

Nitric bath (for desolving more unwanted metals plus for refining silver).

Mix as a SN bath above, then add equal amounts of sulfuric acid (battery acid) to the SN bath.

This will create a weak nitric acid. Nitric acid will not desolve gold, if heated it will dissolve all metals 100% except gold and platinum.

To refine silver in a nitric bath, just desolve the silver and precipitate it back with non-iodized salt.

CHAPTER THREE

REFINING WITH AQUA-REGIA

If you are only after gold, the first optional steps can be skipped.

1-3 ounces of metal per quart of AR will desolve properly without loss. Filter between each pour off with a coffee filter.

First-optional- for platinum metal groups. Precipitate any platinum first with AC (if present), it will drop as a red mud within 30 minutes. Next, pour off the AR and let sit for 24 hours.

any rhodium (rarely present) will drop after the platinum within 24 hours. Do not disturb the AR bath during this time.

Next, pour off into a clean container. Next (going for gold) heat the AR to 120 degrees F slowly add urea, it should fizz (add until it does not fizz any longer). This is neutralizing the nitric acid. Do not be afraid to add extra urea, you must be sure the nitric acid has been put into a neutral state. Next,pour the AR into a clean container, filter if needed.

Add the SMB to 8 ounces of hot water (SMB weight=expected gold return). No more than 1 ounce of SMB per 8 ounces of hot water. Add the SMB/water mixture to the neutralized AR and stir (a foul smell will arise). Add the SMB very slowly, if you have not calculated the amount of gold to expect, you will want to add the SMB/water mixture just a couple ounces at a time. That will require repeating thr precipitation process over and over using your SN detection liquid to check for gold between each dropping.

The gold will drop in 30 minutes to 2 hours as a black/brown mud.

Wash the gold mud with ammonia to purify. The ammonia will remove any trave silver, platinum, palladium and rhodium, pour off the ammonia. Repeat as needed.

Wash the gold mud now with clean distilled water,pour off and repeat as needed.

The gold mud can be tested with SC for purity.

Time to kiln fire the gold mud or torch flame fire on a fire brick for beginners that do not have a kiln.

Kiln fire the gold mud, mix the mud with equal parts anhydrous borax (flux). Place the mix in a crucible and fire at 2000 degrees F until melted. Pour off and let cool.

Refined gold will be .999 fine if done correctly.

SPECIAL NOTES:

Coppe and aluminum can be used to drop the metals from AR. Yeilds a lower level of purity. Use the copper first to drop most of the metals, then the aluminum to drop the copper and any remaining trace elements.

With that said, keep these two base metals out of your AR at all costs.

CHAPTER FOUR

RECOVERY AND REFINING WITH A NITRIC ACID SUBSTITUTE

In a wide mouth gallon jar pour one quart of SN, add 1 to 1.5 cups of water and let it set for 12 hours. Then fill the container three quarters full with sulfuric acid. Let the mixture set for 24 hours, stirring the mixture every 8 hours. The sodium nitrate will not completly desolve.

You now have made a homemade nitric acid that is approximately 70%. Stir daily and just add more sulfuric after you pour off the mixed acid.

Place the material in a clear glass orplastic container. Add the nitric acid substitute, just enough to cover the materials. without heating the bath it will take 10 days to desolve all the non-gold metals.If you heat the bath it will take a couple of hours.

Pour off the bath through a paint filter to catch the gold flakes and pins that are floating. If your bath has real fine gold that escapes through the paint filter, then re-filter with a coffee filter. The waste acid may contain silver, just add nono-iodized salt to the acid. Silver will drop to the bottom as a white mud.

Now the flakes and dust are ready to refine.

CHAPTER FIVE

SILVER REFINING

1-3 ounces per quart of bath will give the best results. Heating the bath to 120 degrees F will speed up the process.

1 pound SN desolved in 1 quart sulfuric acid will make nitric acid.

TIP- SN is hard to desolve, use 1 cup of water for each cup of SN.

once the silver is desolved, wait 30 minutes then pour off into a clean container.

To drop silver use non-iodized table salt, you will begin to see a white mud fall.

Add salt in 15-30 minute intervals until no more silver mud falls. Pour off liquid through a coffee filter and save the mud, rinse with distilled water until clean. Mix clean mud with equal parts soda ash (flux), time to kiln fire.

CHAPTER SIX

ELECTROLITIC RECOVERY AND REFINING

Gold can be recovered and refined at the same time in an electrolitic bath of HA.

C.L.S. Leach Tank-

materials needed-

5 gallon pail

stainless steel cathode plate

3 to 5 inches wide (negative) graphite rod (positive) mounted on the side of the pail is a graphite rod 1-3 inches into the liquid. The other side of the pail has a removable stainless steel plate suspended into the liquid 2-6 inches. Attach a DC voltage souce -/+ as stated. All metals from the raw materials will precipitate and collect on the stainless steel plate as a red/brown mud--brown/black, high gold and pmg's. red mud--copper, lead and silver.

CLS LEACH MIX-

add 1 gallon water to the tank

add 1 gallon of muriatic acid

add 100 ml's of sulfuric acid

Mix 1 ounce CLS mix with a small amount of water, add the CLS/water mixture to the tank slowly. Add water to compensate for evaporation with use. Keep the temperature between 120-180 degrees F.

CHAPTER SEVEN

HOW TO BUILD A MICROWAVE KILN

A microwave kiln is a refractory appliance used inside a microwave oven to focus the microwaves into its interior space. The space is typically occupied by an elevated pedestal upon which the piece or pieces to be melted or fused are placed. It's construction is critical to its proper usage in firing and melting operations. The key to building an effective, safe and controllable microwave kiln operation is in the production of a kiln that focus's microwaves, while simultaneously absorbing reflected microwave energy.

Things you'll need--

2-6 inch glass bowls

spray-on rubber molding release agenreplica stst 101 liquid rubber

ceramic silica casting powder and activator

sharp knife

granular silicon carbide or half inch ceramic fiber board

china teacup

STEP 1- Locate 2 glass bowls of identical size, not to exxced 6 inches across and 3 inches deep. Clear glass is essential to the molding process. Clean them thoroughly.

STEP 2- Spray the intreior surfacesof 1 bowl and the outer surface of the other bowl with spray-on rubber release agent. Turn them upside down and wait 30 minutes.

STEP 3- Pour several ounces of liquid ceramic molding rubber into the bowl with its interior sprayed with release agent. Carefully place the other bowl into the first bowl and allow it to descend to 1/2 inch from the bottom. NOTE: this is an iterative process and it depends on the size and shape of the bowls and the amount of liquid rubber used. Remove and add more liquid rubber as needed. Observe the material through the glass and ensure that no bubbles exist. If there are bubbles, move and twist the upper bowl and recheck. Too much liquid rubber is not a problem; let it over flow.

STEP 4- Remove contact and see if the upper bowl maintains its place in the lower bowl. If it begins to rise, place a light weight, such as a book, on or across the upper bowl to hold it down. Allow the rubber ceramic mold to set for 24 hours.

STEP 5- Remove the upper bowl from the lower bowl. Clean and respray release agent on the outer surface of the upper bowl. Mix 2 cups of ceramic casting powder and its activator, per manufacture's instruction, into a thick and moldable dough-like paste. Press it into the rubber-lined bowl with fingers and work into corners and edges. Press the second bowl back down into the pasted surface and press it firmly, until the paste overflows the edges. Remove the inner bowl and use a sharp knife to cut the flat upper surface on the edge of the material. Cover the ceramic-lined mold and allow it to cure for 24 hours.

STEP 6- Remove the ceramic dish from the rubber mold and scrape or brush any rubber traces from it's exterior. Bake the ceramic dish for 2 hours in a 250 degree F oven. Allow it to cool and bake the ceramic dish a second time at 500 degrees F for 4 hours.

STEP 7- Use any ceramic plate as the base for the kiln and the cast ceramic dish as the shroud or cover. Place a 1/2 inch mat of ceramic fiber board or 1/2 inch bed of granular silicon carbide as the absorbtion base onto the ceramic plate. These materials are dissimilar in price and availability, but there function which is to absorb reflected microwaves, is identical.

CHAPTER EIGHT

HOW TO MAKE AND USE GOLD TESTING SOLUTIONS

Potassium dichromate salts, nitric acid and hydrchloric acids are needed to mix testing solutions. Use the graduate measure for mixing required proportions. In larger cities the PDS and acids are available from chemical supply stores and in smaller cities from local drug stores.

SOLUTIONS: Mark the bottles 1, 2 and 3- do not interchange the caps.

BOTTLE #1- below 14K and base metals

mix 10 grams of PDS with 3/4 ounces of nitric acid and 1'2 ounce of distilled water

BOTTLE #2- 14K-18K

mix 1 part HA to 50 parts NA and 12 parts distilled water

BOTTLE #3- above 14K

aqua-regia mix 1 part NA to 3 parts HA

TO TEST--

File a notch in a test piece, apply a drop from bottle #1- watch for

color reaction

BRASS- dark brown

COPPER- brown

NICKEL- blue

LEAD- yellow

TIN- yellow

SILVER (pure)- bright red

SILVER (.925%)- dark red

SILVER (.800%)- brown

SILVER (.500%)- green

PALLADIUM- none

To determine karat--

File a spot on the test piece and rub it on a test stone to leave a

definite mark. Choose a test needle nearest to the test mark color and rub the needle on the stone next to the mark. If the needle mark reacts sooner than that of the test mark, the piece is of higher karat than the test needle. Trial and error will bring you closer to the correct karat. Use bottle #3 for most of the testing.

TESTING FOR PRECIOUS METALS

SILVER- Pure silver is almost perfectly white, very ductile and malleable. Pure silver is to soft for general use. It is combined with copper to harden it. Sterling silver is 92.5% silver and 7.5% copper. Mexican silver is generally less than 90% pure silver. To test fro silver: file a notch in the piece to be tested and apply a drop from bottle #1. Sterling silver turns cloudy cream. In plated ware, the base metals will turn green.

GOLD- Pure gold is very soft and is combined with other metals to harden it. 24K is pure gold, thus 14K is by weight 14/24 fine gold and the balance an alloy metal. To test for gold: file a notch in the piece to be tested, apply a drop from bottle #1. A bright green reaction indicates gold-plate on copper or brass, a pinkish cream indicates gold-plate on silver, 10K gold will show a slight reaction, over 10K, little or none.

PLATINUM AND PALLADIUM- Use bottle #3, if platinum there will be no change. If palladium, it will turn red after the application of AR (bottle #3). If the reaction to any test is to quick, dilute the solution with distilled water.

CHAPTER NINE

INSTRUCTIONS FOR SCRATCH TESTING GOLD, SILVER AND PLATINUM

CAUTION- Use extreme care in handling gold and silver testing solutions, for they are, corrosive acids. In case of skin contact, flush with large amounts of water. Then treat affected area with sodium bicarbonate or baking soda. If swallowed, contact a physician or hospital at ounce. In case of spills, treat with water then sodium bicarbonate or baking soda.

TESTING FOR GOLD- Scratch thr piece to be tested over the surface of a rough black stone, press well as to leave a visible deposit, preferably a line of one to one-half inches long. For the most accurate testing it is recommended that the user becomes familiar with comparitive testing using standard gold testing needles. For highest sensitivity place a scratch line with a gold test needle next to the scratch line of the metal you are testing. Compare the speed at which the scratches dissolve. If the test scratch dissolves more quickly than the needle scratch, it is a lower karat than your test needle.

Transfer a drop of 10K solution to the scratch made. If the solution dissolves the scratch on the stone, It means the object is less than 10K gold or not gold at all. If the solution leaves the scratch intact, it means the object being tested is 10K or greater.

The scratching and testing is repeated with the 14K solution. If the solution dissolves the scratch on the stone, it means the object is less than 14K gold (if the scratch dissolves slowly and leaves rusty color particles, it is probably 12K gold). If the solution leaves the scratch intact, it means the object is 14K or greater.(Caution:

Many objects are marked 14K, but were fabricated prior to 1982 when it was legal to mark items 14K, but in reality the gold was 13.5K. When testing 13.5K gold, 14K solution will not dissolve the scratch, but it will make it lose its brightness and it will turn into a yellow-rusty color).

The scratching and testing is repeated with the 18K solution and the 22K solution (if available) until the karat of the object is determined. Remember that when the solution being used dissolves the scratch slowly and leaves rusty-colored particles it is probably two karats lower than the solution being used.

On items of heavy weight and volume such as chains, coins, etc, where plating could hide the true metal. It is recommended that a deep notch in the test piece be made and the testing be made with the metal on the inside.

TESTING FOR SILVER- Scratch the piece to be tested over the surface of a rough black stone, press well so as to leave a LARGE and THICK visible deposit, preferably a line of one to one-half inches long.

Transfer a drop of the silver solution to the scratch made. The color reaction of the solution with the metal scratchwill be as follows: (Take into consideration that the background of the test stone is black)

*.925% silver and above, the acid will turn red.

*70-89% silver, the acid will turn green/brown

NOTE: With the silver solution, it is possible to test directly on the piece being tested, however, the solution will dull the polishing of the piece, and leave a mark where the acid was placed.

TESTING FOR PLATINUM AND WHITE GOLD- Scratch the piece to be tested over the surface of a rough black stone, press well as to leave a LARGE and THICK visible deposit, preferably a line of one to one-half inches long.

Transfer a drop of the platinum test solution to the scratch made. (Take into consideration that the background of the test stone is black). If the material on the stone is platinum, it should keep its white, bright color. Platinum test liquid can also be used for 18K and 14K WHITE GOLD. In the caseof 18K, the material on the stone should start changing to a light bronze color in about 3 minutes. For 14K white gold, the material on the stone should disappear in about 15 seconds

CHAPTER TEN

THE 10 MOST PRECIOUS METALS

There are quite a few metals out there that we the people value. Here is a list of the 10 most precious to us, get to know them. Some of these are even more valuable than gold and platinum.

1. RHODIUM

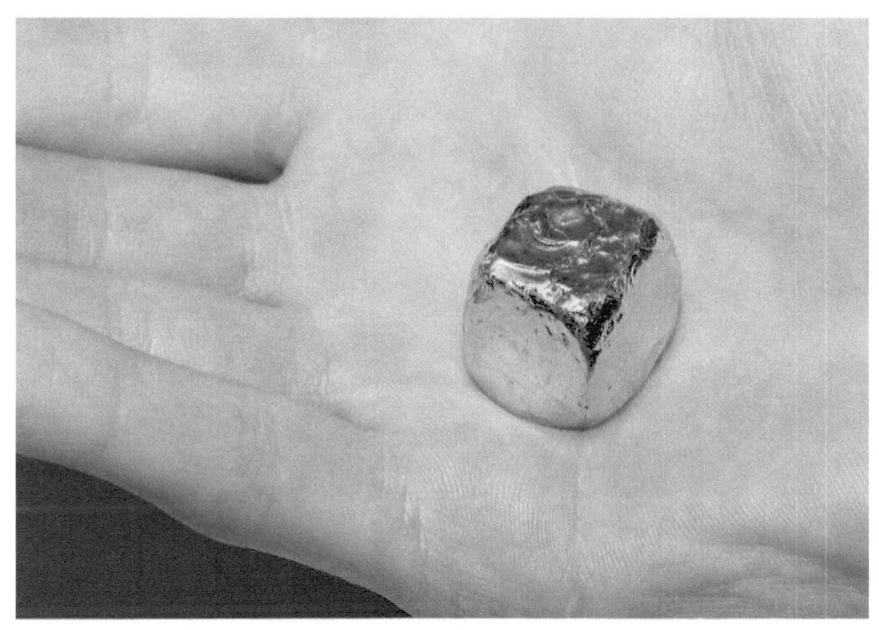

this extermely rare, valuable and silvery-colored metal is commonly used for its reflective properties. It has a high melting point and an amazing ability to withstand corrosion.

2. PLATINUM

platinum has made a name for itself through its malleability,

density and non-corrosive properties. This metalis also similar to

palladium in its ability to withstand great quantities of hydrogen.

3. GOLD

because of its desirability. durability and malleability. gold

remains one of the most popular metals and investment options.

Gold is usually seperated from surrounding rocks and minerals

by minning and panning, upon which the metal is extracted with

a combination of chemical reactions and smelting.

4. RUTHENIUM

this member of the platinum metal retains many of the groups

characteristics, including hardness, rarity and an ability to

withstand outside elements.

5. IRIDIUM

its the most extreme member of the platinum group. this whitish

metal has a super high melting point, is one of the densest

elements around and stands as the most corrosion-resistant

metal. iridium is processed from platinum ore and as a by-

product of nickel mining.

6. OSMIUM

one of the densest elements on earth, osmium is a bluish-silver metal. this very hard, brittle metal has an extremely high melting point.

7. PALLADIUM

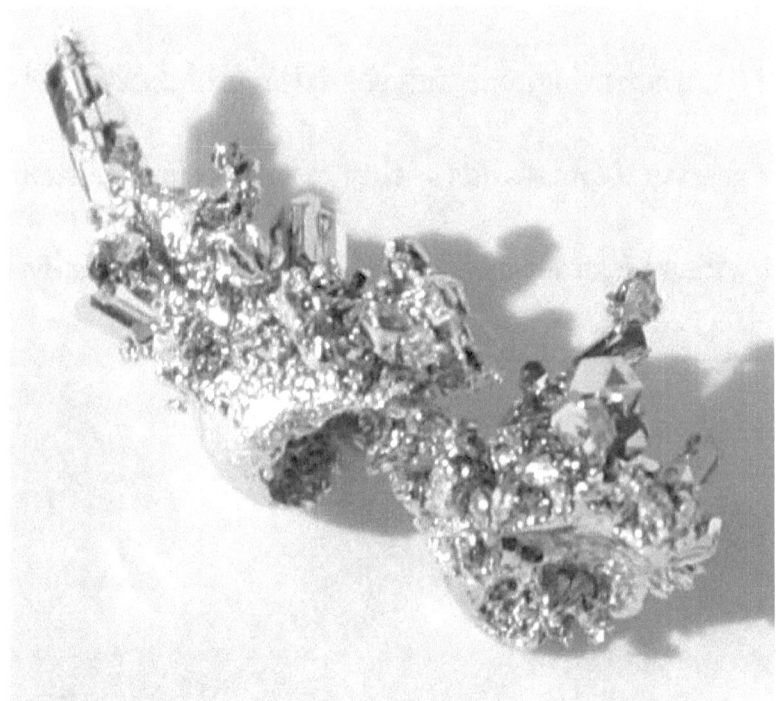

this grayish-white, precious metal is valued because of its rarity, malleability, stability under hot conditions and ability to absorb a considerable amount of hydrogen at room temperature.

8. RHENIUM

one of the densest metals, with the third highest melting point. rhenium is a by-product of molybdenum, which essentially is a by-product of copper mining.

9. SILVER

this element has the best electrical and thermal conductivity, as

well as the lowest contact resistance of all metals.

10. INDIUM

a rare metal produced from zinc-ore processing, as well as lead, iron and copper ores. in its purest form, it presents the color white and its extremely shiny and malleable.

In conclusion I hope this book helps you with your recovery and refining of all your precious metal scraps you find. Although you probably wont get rich any time soon, you will be able to save some for a rainy day.

From my house to yours, peace, love and success.

www.ingramcontent.com/pod-product-compliance
Lightning Source LLC
Chambersburg PA
CBHW021416170526
45164CB00002B/666